Table of Contents

Beekeeping

The Absolute Beginner's Guide on How to Start your First Bees Colony. A Complete Guide on How to Harvest Honey |Helpful Tips for Working With Beeswax.

By OLIVER GOMEZ

Introduction

The technique of maintaining bee colonies to accomplish preferred goals is known as beekeeping, sometimes known as apiculture. The honeybee is the most prevalent domesticated bee species, and stingless bees and other honey-producing bees are saved. Apiculture is permitted anywhere; bees can be kept in towns, farms, ranches, rangelands, deserts, or even woods. City beekeepers may even place their hives on roofs in heavily populated cities!

The bees shouldn't be held, so why do it? Within the agricultural and natural ecosystems, bees are crucial. In addition to producing valuable and healthy products, bees and other pollinators like butterflies pollinate about 75% of the plants that provide 90% of the world's food. A beekeeper or apiarist is the person who maintains the beehives.

Chapter 1: Introduction to Beekeeping

Since the beginning of time, people have consumed honey. It is first obtained from swarms of wild honeybees. As a result, this is known as honey harvesting or hunting. The entire bee colony has to be destroyed to accumulate honey. The natural hives had been broken into, and the bees were suppressed using smoke. Later, people learned that instead of searching for bees in the wild, they could build a hive or residence nearby. Beginning with pottery jars, woven straw baskets, wood bins, and hole logs, they first preserved bees in these materials.

The honey, larvae, eggs, and other components contained in the honeycombs were hammered out during harvesting. After that, a sieve was used to filter the liquid honey. Because every colony was destroyed along with the Queen at harvest, there was no continuity of production. People began to understand bee colonies and biology in the 18th century, which made the creation of mobile hives more appealing. These hives made it possible to collect honey without endangering the colony.

The development of a mobile comb hive by Lorenzo Langstroth throughout the nineteenth century led to a breakthrough in beekeeping. In a rectangular hive box, he created a set of timber frames that allowed a beekeeper to gently remove the honeycombs and retrieve the honey without damaging the combs. After harvesting, the empty combs were returned to the hives for the bees to top off once more. Additionally, people could divide a colony rather than relying solely on spontaneous swarming.

Business beekeeping was made possible by the invention of the mobile comb hive and the wax-comb foundation, including the beginning comb that bees build on and using centrifuges for extracting honey, among other things. The increase in honey production and colony efficiency can be attributed to the prevalence, prevention, and control of bee diseases and the use of artificial insemination of queens and pollen substitutes to maintain healthy colonies. The nineteenth century saw the beginning of commercial beekeeping. Industrial apiarists possess a lot of hives—possibly hundreds of them—that yield a lot of honey.

Chapter 2: Types of Beekeeping

Natural Beekeeping

Herbal beekeeping aims at imitating the bees' natural environment. A beekeeper meets the desires of the bees with techniques and strategies similar to or close to the wild and the herbal nation. The colony survives the winter on honey rather than syrup or artificial nourishment, with little intrusion inside the hive. The bees use their herbal comb in preference to premade or reused comb. Maximum herbal beekeepers do not use chemical treatments in the hive; they use important oils to treat mites and diseases.

Backyard or Urban Beekeeping

Backyard beekeeping refers to maintaining bee colonies in an urban environment. It aims to acquire honey and different bee merchandise by utilizing small-scale colonies. City beekeeping is on the rise in many elements of the sector.

Indoor Beekeeping

It involves keeping honey bee states inside. This might be because of space, observing purposes, or during the slow time of year. Most recent advances have designed some high-level in-house hives that make a permanent place to stay for honey bees to flourish. During winter, a few business beekeepers move their settlements to distribution centers with fixed moistness, light, and temperature. This assists with keeping their provinces solid.

Chapter 3: Legal Obligations of Beekeepers

Beekeeping is a growing hobby, with many people becoming beekeepers in their backyards. Bees are an important part of our ecosystem, and their pollination is vital for our food supply. But keeping bees can come with some legal obligations that beekeepers must be aware of. In addition, there are important steps you should take to protect yourself from lawsuits if your bees become a nuisance to your neighbors.

You can't be a beekeeper without registering. Everyone must register as a beekeeper, even if they keep their bees in their backyard. Registration is compulsory and free, so you don't have an excuse for not doing it.

Registering your apiary is important for several reasons:

- ❖ It helps ensure the safety of people around the colony, including yourself and others that might visit your apiary.

- ❖ Knowing how many colonies you have will help authorities determine whether they need to issue special warnings or take other measures around your area if something goes wrong with your colony. It has to be destroyed or relocated.

If you live in an urban area, you must register your bees with the authorities, which may be a local council, state, federal agency, or a combination. For example, beekeepers in Brisbane will need to register their hives with Brisbane City Council. There are separate registration requirements for the Queensland Environmental Protection Agency (EPA) even further north on the Gold Coast.

Registration is free and usually takes about five minutes to complete online. If you don't do this, it's illegal! In some states and territories, you can also get financial assistance from government agencies if you lose all your bees due to unforeseen circumstances such as disease outbreaks or heavy rains destroying your apiary (bee yard).

The registration includes the bees' location and the owner's identity. If a beekeeper wants to move their hive, they must notify us within two days and describe how it's been moved, including from where to where and by what means.

In addition to these legal obligations, beekeepers should also be aware that:

- ❖ The number of bees allowed in each hive depends on location and time of year.

Registration is important to track the spread of diseases among honeybees. Bees can be very mobile, so knowing who owns them, where they are located, and what diseases they might have is necessary. Registering your bees will help prevent the spread of bee diseases and increase bee safety.

All beekeepers should take their registration duty seriously and comply on time. Registration is important for many reasons, including the following:

- ❖ Registering your hives allows you to track the spread of the disease. This information can help beekeepers and government officials decide how best to address bees in their area and allow people who have had contact with diseased bees to know if they need medical attention.

- ❖ Registered beekeepers receive subsidies from local governments that allow them to buy tools or equipment necessary for beekeeping, such as honey extractors or queen bees. Some states or countries may offer these subsidies only after years spent maintaining an apiary (or beehive).

- ❖ It's also possible that some governments may offer training courses specifically designed for new beekeepers who have not yet completed five years' worth of experience in this field; these courses might teach them how best to manage pests without harming other insects like butterflies or moths at their local farmlands where crops grow naturally without pesticides sprayed onto them daily during daylight hours throughout springtime months only when sunshine reaches at least 80% full strength every hour throughout daytimes which happen every single day between March 21 through October 31 each year so that all plants grow properly before winter arrives again next year around June 1 - September 30 depending upon whether there's enough rain during these periods; otherwise, they'll die out completely until next year when we start over again!

Beekeeping is a great way to make extra money, but it can be dangerous if you don't follow the rules. If you are considering starting a beekeeping business or hobby, you should know about your legal obligations as a beekeeper.

Chapter 4: Learning About Bees

American gold honey (Apis mellifera ligustica) is perhaps the best investment for a farm herd. Large honey producers who tolerate winter well in frigid areas feed gently and have a strong tolerance to illness and natural enemies. In contrast to other types, they seem less susceptible to the connection.

Today, most packaged beehives exclusively sell tested queens and pure Italian hair in packages of two and three pounds. In 1974, such a kit, appropriately called "nuc," was available for between $ 10 and $ 12.00. The businesses operating in this booming industry are in regions with hot weather all year round. They transport bees between April 1 and June 1 (before this time, the weather is too cold, and the summer bees are carried after) when they would otherwise be roasted.

You must prepare your hair before purchasing your bees (wrap the packages together, which I'm sure your post office likes). To put it another way, start putting your beekeeping supplies together once you've ordered your first colony.

Better Know A Bee

Honeybees consist of three different species of bees.:

- Worker
- Drone
- Queen

Each member of the bee colony plays a significant part and carries out a specialized function.

Queen Bee

It can be identified by its abdomen, which is often smooth, lengthy, and far beyond its folded wings.

Role Of the Queen Bee

The hive aims to produce; she typically represents the colony's lone breeding female. Laying starts when workers bring home the first new pollen in the early spring. Until the fall or until there is pollen, egg production continues.

The Queen could sell 2,000 eggs every day at her most successful. The lifespan of a queen bee can reach five years, but it seldom lasts more than two to three years. The elder queens generate too many drones, while the younger ones lay many more eggs. Every year or two, a lot of beekeepers start new colonies. Workers frequently replace older queens without the bees' assistance or knowledge. Even if a skilled beekeeper can produce high-quality queens, novices are typically better off purchasing decent queens from a trustworthy supplier.

What is a Queen Substance?

Queen bees also produce a pheromone known as a queen substance. As food is shared among the bees in the hives, this chemical mixture is transferred from bee to bee one by one. Due to a drop in pheromone levels, workers will become aware of the absence of the queen bee if it is taken from the colony within a few hours.

When a queen is absent, the need to produce a new "emergency" queen from the earliest larvae arises quickly (1-3 days). Additionally, this pheromone prevents workers' ovaries from developing normally. Some people might be lay employees now that the Queen is gone. The workers also rate the number of pheromones the Queen produces. Workers may see it as inferior quality and start planning to replace it if they only receive an insufficient dose daily. To make it simple to locate and verify whether it has been replaced, beekeepers frequently color the Queen's breast.

What Are Worker Bees?

The smallest beehives have the vast majority of workers. Every working woman is a woman, and most cannot procreate. Workers cannot join but can begin fertilizing the eggs that develop into drones in a queen colony without hope.

In the colony, the employees will carry out all necessary chores, including:

- To create a honeycomb, wax from the hive leaks out.
- They collect all the pollen and nectar transported into the hive, turning it into honey.
- Create royal jelly to feed the larvae and the Queen.
- Additionally, they are susceptible to the demands of larvae and queens.
- Mature larval cells are only used for pupation, and the hive is cleaned out of dirt and dead bees.
- Worker bees protect the hives from intruders and maintain ideal conditions by heating, cooling, and ventilating them.

On the sides of their heads, the workers have three basic eyes (ocelli) alongside well-developed complex eyes. To sip nectar from flowers, they have long, well-developed tongues.

Worker Bee Lifespan

Workers in the spring and early summer typically live for five to six weeks. They work in the hive for the first two weeks of their life as domestic bees.

They are out in the wild the rest of the time, searching for food. Workers who mature in the late fall can survive well the following spring. During the winter, a body group should keep the bee queen warm. They will be responsible for raising the first generation of baby bees the next year once laying has resumed.

What Are Drones?

Drones are male bees. The only task of the drone is to fertilize a small sheep.

- ❖ Stronger and larger than the workers.
- ❖ They differ from workers and queens in that they have a little longer antennae and huge, distinctive eyes that connect at the top of their heads.
- ❖ Usually, their mouth parts recede.
- ❖ Drone cells are larger than worker cells when they arise from unfertilized eggs.
- ❖ Drones don't create wax, tend to have pups or gather pollen or nectar. They directly consume the hive's honeycombs or ask the worker bees for sustenance.
- ❖ They are primarily bred in the spring and summer, about four weeks before the new queens emerge, to ensure that plenty of drones are available for the developing queens.
- ❖ They normally divide their day into feeding and rest times and protecting breeding grounds known as drone assembly zones.

As the amount of food available reduces at the end of the summer, drone production will end. Drones are typically driven and guarded by hives by workers before winter. After losing its Queen, a colony can create egg workers that can only make drones. The colony will be all but destroyed if it occurs. Their final attempt to sever the colony's genetic line will be to produce many drones in combination with the virgin Queen of another colony.

Chapter 5: The Year of The Beekeeper

All the driest areas have seasonal honey bees, so the beekeeper's duties follow the calendar. Organizing your beekeeping duties according to the season is possible. Still, it would be best to remember that the seasons in your region may differ somewhat from the dates listed on the calendar. North Dakota's "spring" doesn't begin until after March 21/22, and Florida's "summer" lasts longer than Alaska's ten-week season.

Over the year, keeping an eye on your bees is a good idea. The bees' hive building and daily activity should not be disrupted as much as possible.

Spring Tasks

❖ Beekeeping can begin in early spring when bees are more abundant. Make plans for your hive, buy or build it, and join local beekeeping societies in late winter.

❖ As long as the bees are getting enough food, don't stop. You must feed food for them until flowering flowers are present to offer nectar, as they will have depleted their honey stores over the winter.

❖ Keep a few empty hives on hand in case the bees decide to swarm and need a new home. It's possible to lose bees if you don't follow these instructions. Bees swarm and migrate in the spring.

❖ If a hive is already established, honey can be harvested from the comb when the flowers are blossoming.

❖ Check your hive for a stable brood pattern and replace your Queen if you fear she has died. •

❖ Aim to equalize the number of bees in each of your hives if you have more than one.

Summer Tasks

❖ Your bees will care for themselves during the summer, but you'll need to check on them every few weeks to ensure they're healthy and avoid major concerns.

❖ The bees continuously look for nectar, so you should stop feeding them now.

❖ Make regular checks to ensure that your hives have access to water.

❖ Be on the lookout for stronger hives than others stealing weaker ones.

❖ Make periodic checks to ensure that the Queen is producing enough eggs.

❖ Keep an eye out for Varroa mites.

❖ Make sure your combs hang straight when using foundationless or top bar procedures.

❖ Honey harvest.

Autumn Tasks

❖ During this time of year, honey harvesting is at its peak, and bees prepare for winter.

❖ Harvest honey to ensure that the bees have enough nourishment for the winter.

❖ Look at the combs to see if there are any interesting brood patterns.

❖ Keep an eye out for sickness; cure or discard infected combs.

❖ As long as the bees are disease-free, they can be added to stronger colonies.

❖ Ensure that the hive's entrance is smaller and that mouse guards are in place. Take care of any pest and disease control measures that may be necessary.

❖ After flowering plants and nectar have run out, begin feeding the bees.

❖ Keep the hive from being blown around by the wind during winter, but allow for proper ventilation.

❖ Ensure winter winds don't blow over the hives by putting weights on their tops.

Winter Tasks

❖ You'll help your bee colony settle in and warm up for the long, cold winter ahead before it's too late.

❖ All disease treatments must be completed before a patient can be discharged.

❖ It's important to protect the hives from the wind.

❖ Check your hives often for wind damage, and ensure adequate ventilation as winter approaches. Your bees can handle the cold, but condensation will set in and decimate the bee population if you completely shut the hives.

❖ Keep an eye on the bees in warmer weather to ensure they have enough honey for sustenance. Put in pollen patties or other food when they're low on nectar.

❖ Purchase new beekeeping supplies and equipment. Late February or early March is typically the best time to place these orders to have them delivered in spring.

Chapter 6: What Do You Need to Get Started?

Now we've covered the background of beekeeping and some of the biggest benefits of getting stuck into this hobby, and it's time to look at what you'll need!

As you can imagine, a few key pieces of ki**t are necessary to set up and maintain a beehive**- some you may know, and others you may not. We'll cover a comprehensive list of the basics and a bit of background info about each one.

Let's get started!

The basic equipment beekeeping requires is as follows:

- A Hive
- A Hive Tool
- A Smoker
- Protective Clothing with a Hat & Veil
- Gloves
- Bee Brush
- Feeders

Look at each item in more detail below.

A Hive

One of the most crucial pieces of equipment for housing a bee colony is an actual beehive that your little friends can call their home.

If you weren't aware, hives are a little more complicated than just a wooden box that houses bees; their design involves many different parts and layers to help maximize the efficient harvesting of the bee's products. The hive itself includes everything from the internal "nest" area, honey supers (where excess honey is produced, stored, and harvested by you), an outer cover to protect the hive from bad weather conditions, and a hive stands to keep the bottom of the hive dry and insulated.

A Hive Tool

Hive tools are a kind of metal stick. You can get flat ones but metal ones that are curved at the end, but they all serve a similar purpose. This little stick is handy for various reasons, but the main one is that it lets you pry the hive open. Inside your beehive, you'll find that your bees use a material called Propolis to stick everything together, and it's a resin-type glue. The hive tool allows you to open the hive, scrap Propolis from places, detach comb from the sides of the hive, detach frames, etc. It's incredibly useful.

Smoker

Having a smoker at the ready when working with bees is crucial. Releasing smoke in and around the hive helps calm your bees and make them generally more docile before you open your hive up and dig around. This works because the smoke causes the bees to eat more honey, as they are tricked into thinking there is a potential fire threat nearby, and having a full stomach makes them much calmer and less likely to sting you.

When a beehive faces a potential danger, the "guard bees" act and release a kind of alarm pheromone to warn the rest of the colony. This pheromone smells much like bananas, so if you smell it while you're close to your hive, releasing more smoke to mask it is probably a good idea. Due to this, it's not generally recommended for beekeepers to eat bananas right before tending to their hive, as the bees in the hive might detect the smell and get distressed.

Regarding what fuel to use for your smoker, some common choices are grass, pine straw, or burlap.

Protective Clothing (with a hat & veil)

Now, some people tend to their beehives without any form of protective gear. While this might be comfortable for those more experienced, there is no shame in wearing protective gear. And **if you're a beginner, it's recommended to wear it.**

It should be emphasized that the protective gear you select **should** include face protection because bees tend to go for the face when defending themselves.

Because bees can detect carbon dioxide being released, they tend to react aggressively when mammals breathe out heavily (as they might do if nervous or tired). As people often believe, bees can't sense fear but can sense fear-related behavior, such as heavy breathing. So if you don't feel 100% comfortable around your bees, it's best to keep your face completely covered and protected.

Even if you don't want to purchase a full-body astronaut-looking suit, at least protect your head, face, and hands.

Gloves

As mentioned above, protecting your hands is just as important as protecting your face for several reasons. The main reason you want to keep your hands covered is that while bees can detect nervous behavior through breathing, **they can also detect nervousness through shaky hands.** Of course, navigating your way around a beehive is far easier without gloves, so many more experienced beekeepers choose not to wear them, but **it's highly recommended for beginners.**

Regarding gloves for beekeeping, they're usually made of either soft leather or other durable but flexible materials. The material must be tough enough to shield you from stings but flexible enough to be practical. They also tend to extend the arm pretty far (to just below the elbow) to ensure your hands and wrists are fully covered. If you're shopping around for some, it's a good idea to look for gloves with ventilation around the wrists, as this will be a lifesaver if you're tending to your bees on a hot day.

Bee Brush

A bee brush is useful when you want to gently move your bees off of one comb to a different place. This is necessary for many situations; when you need to harvest your honey, when you need to remove swarms, or repair a broken honeycomb, etc. Using the brush only when you have to is a good idea, as your bees will hate it.

Uncapping Knife/Fork/Scratcher

This is a device you'll need when you're actually harvesting the honey. Uncapping is the process of removing wax that forms on top of honey cells to remove the honey. An uncapping tool will allow you to harvest honey with one simple swipe, simply the process somewhat.

Feeders

A feeder is a gadget with sugar syrup to feed bees throughout the year. This is useful when bees need energy, but honey isn't available for some reason. Some feeders are designed to be put at the hive entrance, but some are designed to be placed inside the hive cavity. Alternatively, you could fashion a makeshift feeder by putting sugar syrup in a plastic food bag and cutting a small slit (around 5") in the top of the bag with a razor knife.

Chapter 7: Building Beehives

You can't be a beekeeper without a place to keep your bees, right? Your job is to provide a suitable habitat for your colony so the bees within it will thrive. Beehives are manufactured with beekeepers in mind, and their design makes it easy for you to observe and handle your bees and collect honey and wax.

The differences between hives and nests are superficial: nests are natural, and humans build beehives. Inside either one, the bees will behave the same. Think of how your needs wouldn't change even if you lived in a cave. You'd have to work harder to get what you need, but your survival would still depend on food, water, warmth, sleep, and shelter. A five-star hotel suite would be more comfortable, but the necessities remain.

Beehives give bees a foundation they can add to and build on, but they'll still be bees within it.

Considerations Before You Set Up

Building beehives are relatively easy, whether you're good with your hands or not. Sometimes, you don't have to build anything; you only have to assemble separate components. Preparing for the beehive, however, demands a little more time and thought. Ask yourself the following questions to ensure your plan is in order.

Where will you place your hive?

This question applies to three accounts. First, you must decide whether you will keep bees indoors or outside. Most beekeepers place their hives outdoors, but in recent times indoor beekeeping has grown in popularity, and hives are made to accommodate it. You'll pay a bit extra for indoor bee boxes, but most people who keep bees in their homes manage smaller colonies, so they invest in fewer boxes. Indoor bee boxes aren't the same as the standard. For your safety, they're encased in glass and function more like a display (a fish tank) than a hive. Not everyone can become a beekeeper indoors, as laws surrounding it are heavily regulated. Indoor beekeeping is just as rewarding and is a good place to start if you're not confident that you could successfully manage a larger colony. However, since it's not the norm, I'll focus on outdoor beekeeping in this book.

If you choose outdoor beekeeping, you must pick the perfect spot to set up your hive. You're looking for a place people are unlikely to visit often. Traffic around your beehives could annoy the bees. Placing your hives in a busy spot also poses a risk to your family, friends, and pets, who might not know how to behave around your bees. On the other hand, you need easy access to your hive and should be able to move around it freely. Shoving your bee boxes into dark corners, which you can't work around, won't do.

Next, you need somewhere flat for two reasons: bees usually build their nests concerning level ground, so if your boxes are skewed, sloping, or misaligned, the honeycomb your bees produce will mimic it. Safety is also a concern here, and if your boxes are unstable, it could spell trouble for you and your bees.

Your hive also needs to benefit the honey bees. It should be close to a water source they can fly to, drink from, and protect from weather like rain and strong winds. Bees are cold-blooded, and their productivity has much to do with the sunshine they get. If your beehive gets direct sunshine in the morning and shade in the afternoon, your bees will get the right amount of light they need for energy without being left to roast or overheat later in the day. To make the most of the morning sunshine, position your hive so its entrance faces south or southeast.

Do you have permission to beekeeping?

Apiary regulation laws vary by state, so you must research this one. I can tell you that Florida has the strictest laws, and registration as a beekeeper there is compulsory. Inquire with your state's agriculture department to ensure you aren't missing any paperwork or licenses. Even if registration isn't mandatory in your region, it's still a wise choice to do so, as your hive will have the state's support. You may even gain access to information that unlicensed beekeepers will find hard to come by, and it will allow you to network. Should you aim to turn your beekeeping hobby into a fully-fledged business, this is a step you shouldn't neglect.

You could face hefty fines if caught with an illegal beehive or colony. The highest fine recorded was $2,000 in 2010, though it's unclear what the violation was (Navarro, 2010).

Beyond licensing, consider the people around you. If your bees will be placed close to your neighbors, it's a courtesy first to find out if they will be okay with it. You never know who has a serious bee allergy, and much as you want to beekeeping, the safety of others should come first.

Then, if you don't own the property you're on, you'll have to get permission from your landlord.

Once everyone who has to know about your bees is in the loop, and you're fully within your rights to set up, you may want to warn your family and friends of your intention, particularly if they visit you often or if children are likely to enter your space. It's a simple precaution but could keep your loved ones safe.

How many bees will you keep?

Formalities aside, you have some planning to do. Before you build your beehive, you should have an idea of how big it's going to be. According to Larry Connor, writing for the American Bee Journal (2015), one Langstroth hive can comfortably house up to 7000 bees. This may seem more than enough to keep you busy, but when starting, you might want to consider building more than one hive for more than one colony.

Many beekeepers have suffered the disappointment of following every step perfectly and then still watching their bees die. Countless factors could bring about the death of a colony—from disease to accident, to mysteries, and all sorts of things beyond your control. Keeping more than one hive means that even if you lose many bees, you won't have to start all over again in your journey. It's also fun to observe and study how different colonies behave and function, so if you're interested in beekeeping from an entomological perspective, multiple hives are an excellent idea.

All this is up to you, though. If you don't foresee enough time or patience to manage multiple hives, starting small is nothing wrong. You will always have the option to expand as you go along. If you bite off more than you can chew, you can pass your bees on to other willing beekeepers, and one is bound to take them off your hands.

Parts of a Beehive

For simplicity's sake, I will teach you about the Langstroth hive. It's by far the most efficient and is the hive style that almost all professional honey beekeepers prefer. They might seem complicated or confusing from a distance, but there are only five basic parts: the covers, frames, supers, deep boxes, and bottom boards. You'll have more choice and variety in how you install or choose to work with each component, but as long as you have all four of them, you'll have a functioning hive.

Covers

The inner cover and the outer cover are the two covers that you can add to your Langstroth. Consider the outer cover to be a building's roof. The remainder of your colony is shielded from rain and sun harm. Although wooden, outer covers may have aluminum flashing or tiles for additional weatherproofing. If you're building your hive from scratch, adding this extra layer will go a long way and save you money in premature replacements. You can install (or construct) a telescoping outer cover.

An inner cover is placed between your hive's outer and topmost hive box. The true purpose of an inner cover is debated, though they're often described as a tool a beekeeper uses to control ventilation, temperature, and humidity within the hive. It seems that inner covers may have been designed with this in mind but have now become a means for beekeepers to manipulate how much room their bees have in the top box. I've also seen talks of how an inner cover makes opening your hive a lot easier, as it offers a place for bees to hang on to and distracts them from your work. I believe the logical use of an inner cover provides your bees with improved circulation throughout your hive. This only applies if the cover has a vent in it. If you're buying your hive pre-built, look out for this feature.

No part of a hive is as debated as the inner cover. Some beekeepers live by them and will argue that the inner cover is the most important component because your entire hive's dynamic will be thrown off balance without one. Other beekeepers claim that they're largely unnecessary, or at the very least flawed, and are tools they can use at their discretion. You must experiment to determine if using an inner cover suits your beekeeping, and I recommend that you give it a go in your early beekeeping days. You can subtract it from the equation if you find it pointless or overcomplicate your work.

Honey supers

Honey supers collect surplus honey that you can harvest without affecting your bees. They come in various sizes, though the most common are shallow and medium. Choosing between the two is a matter of comfort and efficiency. Shallow supers hold less honey but are lighter and easier to lift and work with. Medium supers are heavier but yield a greater harvest. Why don't you get one of each at first, to see which one you like more?

Frames

This is where your bees will build their comb. In Langstroth hives, frames can be extracted and won't cause any damage to the comb or the colony. Frames are the best way to inspect or manage your colony's health. They're also put in place to prevent bees from building comb where they're not supposed to.

Think of them as partitions in your hive that separate each section and keep things organized. Langstroth hives usually have between eight and ten frames and, depending on how you buy or build them, contain pre-coated beeswax foundations. Foundationless frames are gaining popularity as they offer bees a more natural way of building their combs. Note that frames are sometimes called 'supers', so don't confuse them with the honey supers we mentioned above. Frames generally come in sizes shallow, medium, and deep.

Deep Hive Boxes

Otherwise known as deep hive bodies or chambers, this is where your bees will attend to most of their important business. It's the norm to stack two above each other. The lower deep box will be your brood chamber, where the Queen lays eggs and larvae are cared for. If you neglect to install these, the Queen will labor all willy-nilly and might taint your surplus supply with her eggs. The deep upper box is the food chamber, where pollen and honey are stashed away for later use. Don't ever harvest honey from your deep hive boxes. The bees feed from it, and you will harm your colony if you take it away.

Bottom Boards

If the covers of your hive are the roof, and the frames are the walls, the bottom board would be its floor. A bottom board provides a solid and stable foundation for your entire hive, protects your bees from getting too close to the ground, and protects your hive from dampness and dirt. A bottom board can be basic, serving only these two purposes, or it can be screened. Beekeepers are beginning to prefer the latter because they provide extra ventilation. Screened bottom boards also make managing varroa mites (parasites that prey on bees) easier.

Accessories and Extras

Supers, frames, covers, and a bottom board form the necessary structure of a hive, but many beekeepers embellish their Langstroths with some add-ons. The following aren't prerequisites to a functioning hive, though they could improve your beekeeping experience.

- A queen excluder is a frame you can place under your honey super. The gaps of a queen excluder are just large enough for worker bees to travel through easily, not so large that a queen will fit through them.

They keep the Queen bee from getting into your honey super, where she might lay eggs and ruin your harvest.

❖ An entrance reducer decreases bees' passage size to enter and exit their hive and its most important function. You can also use an entrance reducer to control ventilation and air temperature during cold months.

❖ If you like, you can install a hive stand. It gives your bottom board extra elevation. This helps keep your hive dry and clean and also improves circulation around the hive's structure. It's not necessary, but it's a handy accessory that you'll get much use from if you're willing to invest in one.

❖ Bee feeders are optional, but they can improve the well-being of your bees if used correctly. A typical feeder is a small box, jar, stand, or frame placed on or near your hive. The feeder holds syrup—a simple sugar water mix—for bees to tap into when resources are low and a boost is necessary. I'll show you how to make your syrup later in this book. Feeders are meant to mimic your bees' natural foraging and nutritional needs, and they shouldn't be a permanent installation in your hive. The best times to use a feeder are when you're introducing a new colony and the bees haven't had a chance to forage properly; or if you can tell that your bees don't have enough in their honey reserves. Other than that, bees will take care of themselves without help from us humans, and feeding them syrup day in and day out is the equivalent of a person only eating junk food.

There doesn't seem to be a set-in-stone blueprint for a Langstroth hive. Different beekeepers order their hives differently, but there are some norms that you should conform to manage best, inspect and handle your hive. Generally speaking, from the bottom up, a strong hive is fashioned in this order:

❖ Hive stand.
❖ Bottom board.
❖ Entrance reducer.

- ❖ Lower deep hive box (brood chamber).
- ❖ Upper deep hive box (food chamber).
- ❖ Queen excluder.
- ❖ Honey Super (with frames installed).
- ❖ Inner cover.
- ❖ Outer cover.

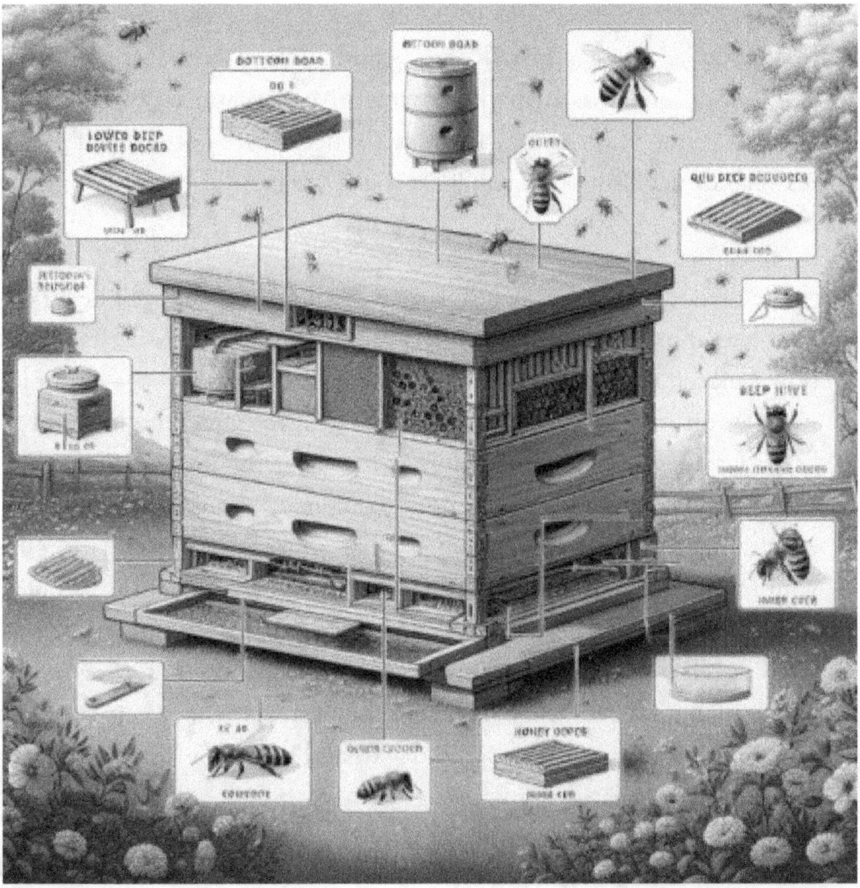

So long as your hive is lifted off the ground, your honey supers are easy to reach, and your Queen can't get into your supply, you will have a hive that you can work with. Here comes the fun part: putting it all together.

Chapter 8: Where Do You Get Bees?

Having all the necessary equipment to set up your beehive is great, but what about the bees? You can use a few different methods to kick-start your colony, and we'll detail a few of the easiest ones below!

Before anything else, we need to answer a crucial question:

When should your beehive be populated?

The spring is the ideal time to repopulate your beehive.

Which will likely begin around April, depending on where you are. To guarantee you start your colony on time, we advise sourcing your bees as early as January because beekeeping is a hobby that has recently gained popularity. There can occasionally be a backlog that takes months to clear.

Please keep reading for our recommended methods to populate your beehive.

Honeybee Packages

A *honeybee package* is a box that contains one isolated inseminated queen bee and around 10,000 worker bees. You can get these packages from bee breeders directly, which are good for easily and effectively populating any hive. Due to their ease of use, these packages tend to sell out pretty fast, so make sure you're well-connected to apiaries (places where bees are kept) within your local area to ensure you can source a package when you need to.

A quick Google search of "bee packages" should do the trick if in doubt.

Just make sure the place you're sourcing from is reliable and practices ethical beekeeping.

How to Install a Honey Bee Package

When you receive your bee package, you'll need to install it within 1 or 2 days. To install it, remove the Queen from the box (keeping her in her cage). You can remove the Queen by first knocking the bee package on the ground to get all the bees to go to the bottom, removing the can of sugar syrup on top, and quickly grabbing the queen bee cage (which should be attached just next to the syrup tin inside the box), and replacing the sugar syrup can as quickly as possible to avoid your bees escaping. The queen bee cage will be sealed with hard candy, which the bees will chew through to allow the Queen to escape. This ritual allows the worker bees time to get used to the Queen's scent and makes it less likely that the colony will reject her. To set the bees off, it is advised that you use a point to poke a hole in the candy. However, it would be best to take great care to avoid damaging or releasing the bees. Place the queen cage at the bottom of the hive cavity if you're working with a top-bar hive (the hive type we will focus on in this guide).

To get the rest of your bees into the hive, tap the package onto the ground again to get all the bees away from the top entrance. Remove the sugar syrup can and swiftly dump the bees into the hive. Try to get as many out of the box as possible, gently shaking it from side to side to make it easier. A few stubborn bees will stay in the package, so once you're happy, lean the package's entrance against the beehive entrance and allow them to make their way inside at their own pace.

Put your bars and roof back onto your hive, and you're set!

Extra: It's a good idea to set up some sugar water solution for your bees when installing them into the hive to help them start building honeycomb.

Nucleus Boxes

A nucleus box, sometimes called a "nuc," is a little beehive. It frequently consists of a wooden box containing three to five frames of honey and brood (bee eggs, larvae, and pupae), one queen bee, and enough worker bees to establish or expand a new hive.

Nuc boxes are readily available from apiaries and bee breeders. Regarding population management, a nucleus box might be preferable to a bee package because the bees will already have some eggs, honey reserves, and larvae on hand. Using a nuc box, they won't have to start from scratch, allowing them to construct the hive much more quickly.

An empty nuc box is an excellent alternative for a temporary swarm trap once you've moved the contents of your old hive into your new one (see next section). The process will be a little bit simpler and less stressful if you catch a swarm in a nuc box since you can leave the swarm there for a time so that it can begin to construct a honeycomb rather than moving it into a new hive right away.

Swarming

When it comes to populating beehives, swarming is the most natural method as it utilizes a naturally occurring phenomenon. Swarming occurs when a colony replaces the old Queen, and the replaced Queen leaves the hive with around half of the hive and some honey. These swarms will land on a structure close to the original hive, and "scout bees" will leave the swarm to find a more suitable location to set up their new permanent hive. During this time, the swarm can be captured and put into a new hive to populate it. These swarms can adapt to a new hive easily and will be ready to start building honeycombs instantly.

Because swarms are local to the area, they'll be found in; this method is *great for ensuring strong genetics amongst the local bee population*. They'll also *probably fare better than bee colonies transported cross-country* (as bee packages or nuc boxes sometimes are). This method is also the cheapest way to populate a beehive, as you can easily capture the hive yourself (read below to find out how to).

So while swarming is a great method for populating hives, keep in mind that swarms **have to be naturally found, so they're not a guaranteed source of bees** (like bee packages are).

How to Catch & Install a Swarm

Before doing anything, ensure you have your beekeeping equipment ready. To catch a swarm, you'll need to have the following:

- a cardboard box (or nucleus box)
- a light-colored bed sheet (or tarp)
- pruning shears
- a bee brush
- tape
- protective gear
- lemongrass oil

- ❖ a ladder or stool (if necessary)

Catching a Swarm

Once you've found a swarm you've determined you can **safely capture**, put on your protective gear, spread out your sheet/tarp under it, and put your box on top of it. Firstly, try putting as much of the swarm into the box as possible (using gloved hands or a bee brush). If your swarm is up on a tree, gently shake the branch to get the bees to fall into the box. If your bees don't stay in the box, likely, the Queen didn't go into the box with them. If this happens, wait a few minutes for the bees to gather around the Queen in a cluster and try again. If the cluster is on a small branch, use your pruning shears to remove the branch from the tree/bush and put it into the box with the swarm. Just be sure to remove all the branches/vegetation before you install the swarm into the hive. If the swarm is on a wall or something trickier to navigate, try misting them lightly with water and brushing them into the box. The water makes it more difficult for them to fly away. If your cluster is on the ground, pour some lemongrass oil into the box and position it on its side so it can crawl into it. The lemongrass oil attracts bees.

Installing a Swarm into a New Hive

Once you've got most of the bees in the box, close it and leave a small opening for any remaining bees or scout bees to enter. Leave the box until the sun sets to ensure scout bees have enough time to return from their rounds. Close the box completely using tape at nighttime and wrap the swarm in the bed sheet/tarp. Make sure there are air holes. Be careful when transporting and installing the swarm box into their new hive the following morning. Once you are ready, open the box and shake the bees into the hive. Put the bars and lid back on, and you're done!

To encourage swarming, you can set up swarm traps. These work best when they are put in a high place, like up in a tree. Putting a few drops of lemongrass oil in these traps will encourage swarms to enter them.

Bait & Trap

Baiting and trapping is a method that is an extension of swarming, so be sure to read the above info on swarming before reading this part. As you know from before, when swarms form on their "temporary" new base, the scout bees will leave the swarm in search of a new appropriate hive location. Once they find one, they return to the hive and direct the bees to the new place using a special dance (affectionately called the waggle dance). A few worker bees go to the new location and release a pheromone to make it easier for the other bees to follow them, a pheromone that smells exactly like, you guessed it, lemongrass oil. As mentioned before, lemongrass oil is a great lure to draw bees to a location. So to draw bees into a swarm trap put a little drop of the oil (you don't need to use too much!) inside and let the oil work its magic.

Splitting

Hive splitting is where a beekeeper "splits" a pre-existing hive into two by removing its frames (or top bars) and moving them into a new hive. On these frames/top bars, there should be unhatched eggs, nurse bees, and honey from the previous colony. If the bar/frames you remove don't have a queen bee- you don't need to worry too much, as the worker bees can raise a new one by feeding an unhatched egg royal jelly. This process cannot happen if the egg has already hatched, so ensuring your frames/bars include unhatched eggs is important. You can, of course, buy a queen bee and add them separately, but most beekeepers tend to support the practice of queens being naturally added to a hive instead. The practices used to inseminate queen bees can be problematic, so research before buying a queen bee to see if they have been ethically raised.

To add a "split" to your new hive, remove a frame/bar from one hive and install it into a new one. It's very, very simple.

Splitting is another popular choice for populating new hives these days, as it's a pretty easy and natural solution and makes the most of a bee colony you already have! It's also completely free, as you don't need anything more than an existing colony. If you're new to beekeeping, why not try to make some beekeeper friends and strike a deal with them to split one of their existing hives?

Chapter 9: Safety First! How to Harvest Safely

This chapter will cover what you must wear when handling your bees and what to do when you get stung. Unfortunately, getting stung is a part of having a beehive, and knowing what to do when you're stung won't be an as traumatic experience as if you stepped on a bee in your yard.

Safety Kit

Think about your attire as a simple approach to prevent your bees from acting defensively in the first place. Beekeepers know strong colors like black make the bees aggressive, so they dress in light, silky textiles. Fluffy materials like flannel, sweaters, and athletic socks conveniently capture bees' hooks on their feet. On hotter days, wearing some light, breathable Clothing is preferred because your sweat will annoy the bees.

Bees are also sensitive to movement and light. When handling them, take off any jewelry. Any jewelry that is sparkling in the sunshine may irritate them. Any long hair should be pulled back or tucked in to keep it out of your way and from moving about in the wind.

Because they are clever, your bees will use a gap of no more than three millimeters to get inside garments. Tuck your jeans into your socks or boots to prevent them from climbing up the legs of your pants. Ensure that any gaps are filled. You might also tape your pants or sleeves with some elastic bands.

Pheromones and fragrances are vital to a bee colony, making them a simple approach to elicit a protective reaction. Avoid using fragrances, hair products, deodorant, or after-shave on your way to the colony. For instance, you might want to refrain from gassing up your car on the way to the hive.

How to Treat Bee Stings

You'll get stung no matter what you do, and it would be best if you got ready in case something like this happens. You will be close to your bees and stealing their food, and they will sting you regardless of how much planning and research you do or how much protective gear you wear.

Beekeeping is probably not a good idea for you if you are aware of your strong sting allergy. If you truly want fresh honey, buy some from your next-door neighbor.

If you're concerned about being stung by bees, calm yourself and remember that most beekeepers become used to being stung and that the stings grow less painful with repeated exposure. Stung beekeepers will produce antibodies to the venom, and to maintain these antibodies, some beekeepers advise getting stung each season. But make sure you take all the appropriate safety measures.

To one or many stings, each person responds differently. Some patients experience localized responses, like welts and swelling, while others have systemic reactions, such as anaphylactic shock, when breathing becomes challenging. It is preferable for a beekeeper to always have a sting kit on hand, which calls for a prescription. Epipens and antihistamines are part of this. Ensure you understand how to use the equipment in an emergency.

What You Should Do When You're Stung

The first thing you must do after being stung is to maintain your composure. Scrape out the stinger, exhale a little smoke close to the sting site, and then step back from your hive for a while. Then, once your hive has quieted down, venture back outside!

Scraping off the stinger is preferable to squeezing it since it reduces the amount of venom injected into you. The majority of folks squeeze the stinger before pulling it out. A complete dose of venom is released into your skin when you pinch it, which also pinches the stinger pouch. The amount of venom discharged into you will be reduced if you scrape the stinger using a credit card or the hive tool. As a result, the sting has been lessened and won't hurt as much.

Because the sting will prompt additional bees to sting, you should blow smoke near where you were stung. The smoke diminishes the first sting's warning scent. It's a good idea to leave your hive for a few minutes—not so much to let the bees calm down, but rather to calm your anxiety and ensure everything is okay before you return to working with your hive.

Determine whether the weather is to blame if you've been stung. Avoid tending to your bees on overcast, chilly, or windy days. It is a sure way to get stung to work them before a storm. On days with lousy weather, they are simply irritable.

Tips for Reducing Stings

1. You should choose a good day. Good days are sunny, warm, high pressure, and not very windy.
2. Pick a good time, such as between ten in the morning to two in the afternoon. Working them before ten in the morning or after two in the afternoon will likely result in a sting because the hive has more bees. The hive is smaller and easier to control between ten and two because the foragers are out.
3. Avoid using perfume, strong soap, or hairspray when you're near your hive. It may soon turn ugly if you get all dolled up and scented to work a hive. Bad body odor will also cause some stings, though. Be clear as a result, but don't go overboard.

4. Avoid donning dark apparel. Bees like to sting in shadowy places. They avoid white. If you don't wear boots, white Clothing is preferred, especially white socks.
5. Never swat a bee that is swarming near you. They don't disappear with swatting; it has the opposite impact. When swatted, they become more aggressive, and if you try to flee, they follow you. You cannot outrun a bee since its maximum flight speed is eighteen miles per hour. So, keep calm and watch to see if the bee disappears. Guard bees are typically the first to buzz your face and emit a louder-than-normal buzzing noise to scare you. Don't be fooled by their intimidating buzzing; the stinger is the same size as any other bee's, and it won't hurt any more or less. If you remain still, the bee may become weary of frightening you and fly away.
6. Do not exhale forcefully or sweat directly onto your bees. Inhaling regularly is preferable to hold your breath, but you should avoid breathing too close to the camera. Leaning slightly to the side of the hive will prevent you from dripping sweat onto the bees on a hot, muggy day.
7. The nectar flow is when bees are the calmest, but if you disturb the hive during the nectar flow, they won't be as productive. Nevertheless, when the flow is in motion, they are the calmest. Numerous floral sources are available in a flow, and production is abundant.
8. Frequently utilize your smoker. The beehive has to be smoked, be kind to them, and don't smoke too much to avoid upsetting them. When relaxing in the hive, a little smoke goes a long way. But don't handle the bees without smoking them first.
9. Use calm, soft motions. Avoid jerky motions, and never accidentally drop a frame or hive tool upon your bees. Bees are incredibly sensitive to vibrations despite not having ears. You should back away gently, remain motionless if you're not being pursued, smoke, and try to put things back together when the bees calm down if you drop something or tip the hive over.
10. Always keep your hot and veil on. You may not care if you get stung below the neck, but you can't take the chance of getting stung in the face or eye. If you don't cover up and wear a hat, you'll regret it, and if you want to prevent getting stung, you should put on protective gear and tape all the gaps in your Clothing.

Normal and Life-Threatening Reactions: Their Differences

Because of the swelling, itching, and changes in skin color, some claim to be allergic to bee stings. A bee sting causes everyone to react in this manner. Only individuals with an established immunity from several stings do not swell up and itch. You'll probably stop turning red and swelling up by the end of the season. The area where you were stung will grow red and swell up after the winter until your immunity returns. The enlargement might move. The edema that results from an arm sting typically extends a few inches from the sting location. This response is typical. After a while, the swelling will decrease, and the itching will disappear. The swelling could not completely go away for up to a week. This does not imply you are experiencing a potentially fatal allergic reaction to your bees. It signifies the poison from bees has caused your body to respond normally.

When your body enters anaphylactic shock, reactions become life-threatening. Though this is a rare reaction to a honey bee sting, it must be treated seriously, and the victim must seek medical assistance immediately. Breathing problems and other systemic problems will result from this reaction, and a medical specialist swiftly fixes it, but timing is crucial.

For everyone else, a bee sting causes temporary pain, followed by swelling, redness, itching, and pain for a few days if they are not sensitive to bee stings. That is a common response. Because of this, beekeepers have produced friendly bees. Because of this, you ought to start with an insect known as the Italian honey bee.

You'd have to go into anaphylactic shock to die from a honey bee sting because studies have shown that the average person weighing 160 pounds can sustain 1,500 stings at once and still live.

Male bees, commonly called drones, can't sting since they lack a stinger. Catching a male bee, posing as a worker bee, and wearing it on your face will make you look incredibly cool in front of your buddies at a backyard barbecue.

Hives can, regrettably, become aggressive. Queens who are combative will have combative daughters. When small animals upset the hive at night, the bees become irritable during the day. They spent the entire night attempting to get rid of a raccoon, and after that, you'd also be quite cruel.

It's time to replace the existing Queen when a hive becomes too hostile to tolerate. The situation should be more peaceful after four weeks when all the bees are from the stock of the new Queen. Avoid putting up with a nasty hive queen. Requeen all at once so you can take pleasure in beekeeping.

Chapter 10: Harvesting Honey and Beeswax

The honey is ready now! You have completed all the necessary tasks to keep your colony healthy and produce honey. It's finally time for some tasty goodies! Additionally, homemade honey tastes much better and is much healthier than store-bought honey. How to go about collecting and extracting your honey is covered in this section.

Just be aware that your honey harvest will be modest if you're starting beekeeping and it's your new colony's first season. This is so that new colonies have a whole season to establish a sufficient number to enable them to produce surplus honey.

Before continuing, it is important to remember that you can collect two primary kinds of honey: extracted honey and comb honey.

Extracted honey is the most typical kind of honey. Using centrifugal force, extracted honey is taken out of the honeycomb (by using an extractor). The liquid is subsequently put into bottles.

Comb honey involves using the comb to collect the honey. Comb honey, which consists of both the wax and the honey, is completely palatable.

However, I would continue gathering extracted honey for beginning beekeepers. This is because to harvest comb honey, optimal conditions are necessary, including strong, highly productive hive populations, ideal weather, and an abundant nectar supply. Typically, this may be saved for those with more experience.

As a result, we'll start by discussing how to gather extracted honey.

1. Timing

When the hive is full of capped, cured honey, it is when honey is typically harvested. I would advise checking on your honey supers sometimes to see how your young bees are doing. It's time to start harvesting if 80 percent or more of your honey suppers contain sealed (capped) honey.

You can see capped honey in the top two-thirds of the image in the previous photograph. Uncapped honey is contained in the bottom third of the cells. If the honey is properly cured, it can also be removed. Turn the frame upside down and gently shake it to check if the honey has been cured. It is not yet cured and should not be harvested if honey is seen dripping from the cells. Your crop will be ruined by uncured "honey," which is not yet honey. Put the honey super back into the hive and wait for your bees to cure the remaining honey cells.

It's crucial to avoid waiting too long when scheduling your honey harvest. Your bees will start eating the honey once the final significant nectar flow has passed, so much of the honey you hoped to harvest will be lost. The temperature is an additional consideration. The weather may be too chilly to harvest honey if you wait too long, and honey thickens when it becomes cold, which makes harvesting impossible.

Beginning in your second season, you will typically perform your first harvest in the height of summer, typically around late May, and a second one in the middle of September, before the weather turns chilly. These projections, however, could change from year to year and also based on where you live. To get your bees through the winter, leave them 60 to 70 pounds of honey at your second harvest. The last significant nectar flow is about to cease, so if this is your colony's first season, you might only get one harvest.

1. Removing Honey Supers From the Hive

Honey can be extracted in a wide variety of ways. The best and safest methods of extracting honey will be covered in this section. Wearing your veil and bee-resistant Clothing is necessary before harvesting, though. Wearing protective goggles will also go a long way at this time because your bees will be at their most defensive. You'll also need a smoker, an empty honey super, and an old towel.

1. Use your smoker liberally to quiet your bees, and then open your hive as you would for a routine inspection.
2. Bees should be removed from honey supers. Take one frame at a time and shake the bees off to accomplish this. You can alternatively use a bee brush to remove your bees from the frames. To protect any bees, make careful only to brush upward. Alternatively, you can use a fume board and bee repellant to eliminate the bees from the honey dinners.
3. Place the cleared frames into the empty super after removing the bees.
4. Cover the super with an old towel to stop your bees from stealing the honey.
5. Take the honey dinners out of the hive and place them in an area free of bees.

1. Extracting Honey

As soon as the supers are removed from the hive, you should begin to extract your honey. So once inside, it's time to begin harvesting your priceless honey.

What You Need:

❖ Honey strainer

❖ Uncapping knife, fork, or scratcher

❖ Bottling bucket

❖ Honey jars

❖ Honey extractor – A cylindrical piece of equipment is a honey extractor (see picture below). An excellent extractor may be bought for anywhere between $200 and $500 and higher, and they come in various sizes and types. Budget plastic extractors are more affordable, though I wouldn't suggest using them.

The steps for removing the honey from your frames are listed below:

1. Use an uncapping knife, fork, or scratcher to remove the wax-sealed comb cells from both sides of the frame. I prefer to first expose most honey cells using the uncapping knife before uncapping any cells I may have missed with the uncapping fork.
2. In your honey extractor, position the uncapped frame vertically. Start spinning the frames once you have removed the caps from enough frames to fill your extractor. Honey will dribble to the ground, and the wax comb can be damaged if the frames are spun too quickly.
3. Put the frames back into the honey super once both sides have been emptied.
4. Continue with the subsequent frames.
5. At the bottom of your extractor, there is a valve that you should open. By doing this, the honey can pass through the stranger and into your bottling pail (see picture below).
6. Put your honey in tightly closed bottles!

You're done now! The results of your labor can now be enjoyed. But first, they must be cleaned before putting the honey supers away for the winter. The honey supers should be placed on top of your hive (between the top deep hive body and the inner and outer covers) and left there for a few days to help clean your extracted frames. Your frames will be dry, clean, and prepared to be stored for the upcoming honey season once your bees completely suck off any remaining honey.

The last step is to treat your honey supers with wax moth control before storing them for the winter. You can do this by freezing the combs for at least 24 hours or using PDB crystals to treat the supers.

1. Harvesting Beeswax

The cappings you cut off when extracting the honey are your primary wax crop for the year.

What You Need:

- ❖ Block mold
- ❖ Cheesecloth
- ❖ A honey strainer or colander
- ❖ Double boiler

Steps to Follow:

1. Allow as much honey as possible to drain off your cappings by letting them drain for a few days.
2. Use warm water to rinse away any residual honey from the cuppings.
3. Through a honey sieve or a colander, drain the cappings.
4. Put the cleaned cappings in a double boiler to melt the wax.
5. To get rid of any contaminants, strain the melted beeswax using cheesecloth.
6. Before removing the wax from the mold, let the molten wax cool and solidify. Use a mold or an old cardboard box, like a milk carton.

Make the most of your crop now that the harvesting season is finished!

Chapter 11: Beehive Care and Maintenance

Here are some general tips and guidance for beehive care and maintenance. Taking care of your colony and keeping your hive healthy go hand in hand, and bees that live in a healthy hive are content.

First off, while the quantity of upkeep required for your hive can change, it's crucial to maintain some consistency during this period. You will get more out of it if you put more effort into it, just like with any abilities and interests.

Your bees will create considerably higher-quality materials if you keep up with beekeeping properly. Additionally, your bees will be much healthier, happier, and more productive (and they won't be as prone to swarm).

Most specialists concur that the best honey harvest occurs in the spring when beehives need the most upkeep.

Steps To Maintain Your Hive

A beekeeper must follow some steps to maintain their hive:

1. General Reparations

Your beehive will gradually lose strength and become less livable for your bees. The weather, which can result in problems like warping, rotting, cracking, and rusting, is the main cause of this. Larger animals and insects can potentially harm your hive.

Your hive needs to repair any visibly damaged pieces as soon as possible. If they can no longer be fixed, they must be replaced.

You can ensure your hive can withstand the environment and other harmful circumstances by frequently replacing and repairing damaged pieces. Additionally, it will keep your hive looking good!

2. Entrance Blocking

It is only really required to use entry barriers during the winter. Beekeepers restrict the number of hive entrances to trap heat inside and reduce the need for bees to heat the hive. Additionally, it keeps the ventilation at an adequate level and prevents the entry of other undesirable objects.

Your entrance may be blocked by:

- Using an entrance reducer

❖ Erecting a mesh or other perforated material with holes above the entrance that are large enough for your bees to pass through.

❖ Putting pieces of wood in some of the numerous entrances to the hive

3. Excluders and Supers

The queen excluder, a layer that separates the brood from the comb, stops the Queen from laying eggs in inappropriate places. When their honeycomb or brood combs are full, bees fill the space in supers with honey.

Using a queen excluder makes it much simpler to maintain your hive, so it will continue to function well. If you don't already have one, install one.

Aside from keeping your hive functioning properly, adding more supers will provide your bees with the extra storage they require during their busiest season (spring to summer).

4. Replace Lost Beeswax

Bees use beeswax to line the inside of the hive to strengthen and improve comfort and create combs to hold their honey. It is, therefore, undesirable if any of it is damaged or lost.

A beekeeper might accidentally break off beeswax when opening their hive or removing frames, and it happens rather frequently, and beeswax will be lost when honey is obtained. To avoid doing this, use a hive tool, but make sure you also attempt, if at all possible, to repair any broken wax.

You can buy natural replacement beeswax from a store or utilize extra wax from another hive if you need to replace huge chunks of wax.

5. Sufficient Ventilation

The hive's interior may suffer serious harm if over or under-ventilated. If the hive is not properly ventilated, water or even ice may develop inside, endangering the hive's structure and the bees. Inadequate ventilation can also lead to the hive overheating, damaging the structure and making the bees uncomfortable. You'll see your bees engaging in particular actions, including bearding (see image) or fanning when overheating.

6. Windbreaks

Beehives can suffer great damage from strong winds. In addition, pushing damp or chilly air into the hive's cavity can overturn the entire structure, wiping out most of your bees' laborious efforts.

Think about adding a windbreaker to protect your hive from powerful winds. There are options made by humans and by nature (trees, shrubs, etc.).

Find out where the wind usually comes from in your neighborhood.

7. Rain Protection/Shade

Your hive might have major weather-related issues. The hive can be severely harmed by rainwater or any other type of moisture, which can also kill your bees or substantially upset them. The same unfortunate consequence can be caused by excessive direct sunshine.

- It's a good idea to use an outer or slanted cover to shield your hive from the rain so that the rain will roll off it. It would be ideal if it were made of plastic or metal.

- You can either make something to create shade or locate your hive in an area that will minimize direct light exposure.

If you decide to use a rain cover, be careful to take it off sometimes because doing so can hinder the hive's ability to ventilate properly. If your hive does unavoidably get too much water in it, set it up on a slight incline so that it can drain.

8. Woodlice and Termite Protection

Beekeepers live in constant fear of pests like termites and woodlice. These critters can readily feed on your hive and gradually kill it because most of us choose to build our beehives out of wood. Termites and woodlice can easily hide inside rotten or old wood, and they are most vicious during the drier months.

- You can use baits, liquid pesticides, or soil treatments to reduce the number of these pests. Please find the best solutions from your local beekeepers, and before using them, make sure your bees won't be harmed.

- You can also stop them by clearing the vegetation around your apiary.

9. Predator Protection

Several creatures, such as bears, mice, skunks, raccoons, etc., will likely pose problems for your hive. Let's examine each of these in greater depth and offer possible fixes.

- ❖ Bears enjoy eating honeycomb and destroying hives. If bears are a problem where you live, you can keep them away by enclosing your hive in a sturdy fence or boundary.
- ❖ Mice are notorious for constructing nests in beehives, which can lead to various issues for you and your bees. You can deter mice by positioning a mouse guard beside your hive's entrance.
- ❖ In addition to digging holes in hives, mice, and skunks frequently do so, destroying the comb and rendering the hive inhospitable. Making sure your hive is on a stand that is appropriately raised is the greatest approach to deter skunks.
- ❖ Your beehive may sustain significant damage from raccoons as well. You may stop this by adding weight or lock to your hive's lid or roof because they frequently enter beehives through the lid to get inside.

Keep Things Clean

The final guidance I'll give you is to keep things clean. The saying "Cleanliness is next to godliness" is one you may be familiar with. In a beehive, it is undoubtedly true. A healthy beehive places great importance on cleanliness, which is very significant. In actuality, the beekeeper can transmit many infections that harm beehives.

Nearly all of your beekeeping equipment needs to be cleaned frequently, and you can clean most of the equipment with hot water and soap. Baking soda, vinegar, and washing soda are some of my preferred natural cleaners, and I only use bleach when necessary.

- ❖ **Hive tools:** Use vinegar or bleach with water in a spray bottle or bucket to clean your hive tools. Do not combine the two solutions.

Your cleaning supplies can be sprayed, wiped, soaked, then dried, and your bee brush is part of this.

❖ **Bee feeder:** Thoroughly rinse after washing in warm, soapy water. Each time you add new bee food, follow these steps. Clean your feeder with a bleach solution if your colony is ill.

❖ **Water dishes:** Remember to thoroughly rinse after washing in warm, soapy water. At least once per week, carry this out. Apply a bleach solution to your feeder to clean it if your colony is unwell.

❖ **Bee suit, gloves, and veil:** There should be a label with washing instructions on your bee suit and gloves. I always hand wash the veil in the sink using warm, soapy water, rinse it, and then hang it to dry.

❖ **Beehive and frames:** You must wash your hive before removing a frame or box, changing a board, or harvesting honey, especially if a hive was lost to illness or disease. It is crucial to wash the hive before adding bees if you purchased a secondhand one or received one as a gift. A large bucket or tub should contain one cup of bleach for five gallons of warm water. To remove any debris, use a wire scrub brush. Everything should spend 15 minutes soaking in the bleach water. Repeat the soaking, remove, and scrub if necessary. After cleaning everything, you can store it in a sealed plastic bag for protection or allow everything to dry completely before putting it back in the hive.

❖ **Honey extraction equipment:** Use warm, soapy water and washing soda to clean all honey extraction machinery. Make sure to remove the honey from your tools completely. Before storing it, could you give it a good rinse and dry it?

❖ **Smoker:** Just like your fireplace chimney, keeping your smoker clean and clear of buildup is essential. Your smoker's walls should be scraped using a wire brush, eliminating stray materials. Spray the remaining interior surfaces of the smoker with vinegar, then sprinkle baking soda on top. Then do it again. Remember to avoid getting your smoker's pump wet. The best ones I've found are tiny wire brushes that are roughly toothbrush size. When you finish using your smoker, or as soon as you see any black accumulation inside, clean it.

Sanitary beekeeping techniques will help shield your hive from transmitting illness and disease. Do not attempt to clean this yourself if you have AFB or EFB, though. Inquire with the county extension. They will instruct you on disinfecting every piece of equipment and may even demand that you burn your hive or frames.

Chapter 12: Bee Health - Pests And Diseases

Many enemies often attack honey bees, that is, pests and diseases. Usually, the most common or dangerous enemies require your attention as a keeper to get the best out of your beekeeping adventure. If you cannot detect and prevent pests and diseases associated with bees in time, this might cause harm to your honey, wax, other bee products, and the entire colony.

Let's look at a few of these pests and diseases:

Predatory Wasps: Vespa velutina (V. Auraria). This pest builds nests on top of structures and trees.

Nature of damage - The wasps capture and kill the bees at the hive entrance and may even succeed in wiping out the weak colonies.

Prevention and Control - By flapping or burning the nests at night, you can eliminate the fertile females who come to the apiary in the spring to lay their eggs. However, in areas with fire risk, destroy the nests by spraying them with a potent insecticide.

Wax Moth, Galleria melonella

The comb's midribs were tunneled through by wax moth larvae, which caused a nature and degree of damage. When there is a serious infestation, all further brood rearing is halted.

Prevention and control - All fissures and gaps should be sealed. Minimize hive entry

Control in storage

Burning sulfur at 180 grams per cubic meter will clean the stack (fumigation Sulphur fumes). Naphthalene flakes should be placed in moth-proof stacks after fumigation.

Ectoparasitic Mites

Nature of damage: Only bee broods are consumed by this mite, and dead broods are flung outside of the hive by workers if there is a serious infection of this mite. If control measures are not implemented, bee colonies may leave the area.

Control

The control methods depend on the type of destructor you are dealing with because two forms of these mites attack bee hives.

- Tropilaelaps create 200 mg/frame of sulfur dusting on the tip bars.

- Varroa destructor: Fumigation using formic acid at a rate of 5 ml per hive on sponge pads protected by perforated plastic bags.

By applying sugar (finely powdered sugar) at a rate of 30g per frame and then sweeping sugar down between the frame spaces with a bee brush, the degree of mite infestation can be kept to a minimum.

Honeybee Diseases

Numerous diseases, including those brought on by viruses, bacteria, and protozoans, harm honey bees. These endoparasitic and ectoparasitic mites and viral, protozoal, and bacterial pathogens attack honey bees. Some examples of these ailments are:

Viral Diseases Such As Sac Brood Disease

It spreads into the tissues of immature larvae and multiplies there. These larvae typically don't pupate and stay sprawled out on their backs with their heads pointing toward the cell caps.

Management Of Viral Diseases

- Adults in unhealthy colonies should be moved to a new, clean hive with a comb foundation.

- Feed them more often with sugar syrup and extra pollen.

- To stop future contamination in situations of severe infection, burn the combs holding infected larvae.

- By taking a break from raising the brood by de-queening or enclosing the Queen, the bees will be motivated to remove diseased, dead brood quickly, which will help keep the infection under control.

Bacterial diseases such as American Foul Brood (AFB) and European Foul Brood (EFB)

An uneven "paper box" of uncapped, pierced, or sunken capping separates the infected and healthy brood. The sick brood starts looking dull white before turning bright or dark brown.

Management

In regions with abundant nectar and pollen flow during the active season, honeybee colonies could be established.

Infected colonies with AFB used artificial swarming or "shook swarm" techniques during the post-honey flow seasons. This method destroys infected brood combs after adult bees have been moved to a healthy hive.

By fumigating with formaldehyde, the combs and equipment can be sterilized. When administered with strong sugar syrup, sodium sulfathiazole (1.5g/15l) and oxytetracycline hydrochloride (0.4g in 5l) inhibit the disease.

Yeast infections like Chalk Brood and Stone Brood

Symptoms

- ❖ Pre-pupae and larvae of the fungus are typically found on the periphery.
- ❖ After cell capping, the infected larvae pass away and turn white. Grey and black coloration then develops on the formation of fruiting bodies.
- ❖ Larvae swell and are overgrown by fluffy cells like mycelia.
- ❖ The diseased larva dries out, becoming a hard, skinny white chalk mummy.

Management

- ❖ the combining of adult bees with brood combs to strengthen weak colonies.
- ❖ Regular replacement of worn-out combs with new ones.
- ❖ Using methyl bromide and ethylene oxide to fumigate hive tools and combs.
- ❖ Chalkbrood can be controlled by dissolving trichloro iso-cyanuric acid (TCA) in water.

Protozoan Diseases Such As Nosemosis.

It is an obligatory parasite that grows in the adult bees' stomach tissues and is known to reduce honeybee lifespan.

Symptoms

Infected colonies' bees exhibit restlessness.

Infected bees' hind wings may separate from their fore wings.

Infected nursing bees don't make enough royal jelly.

The newly emerged adult bees with the pathogen's hypo-pharyngeal glands do not fully mature and finally atrophy.

Management

- ❖ keeping bee colonies healthy, with a productive queen and enough food supplies.
- ❖ You can switch out your old combs with new ones to avoid spreading germs.
- ❖ The Nosema spores would be inactive by acetic acid fumes (60 percent).
- ❖ When fed to bees at 0.5–3 mg/100 ml sugar syrup dosages, the antibiotic fumagillin reduces Nosema infection.

Chapter 13: Getting the Best Price For Your Honey

A beekeeper must comprehend what elements can affect the quality of the honey and how to maintain the highest value possible, as the quality of your honey will determine a major portion of its value.

The bees will largely take care of maintaining the hive's optimal condition while the honey is still inside of it. The honey will eventually get darker if kept in the hive for a long time. Additionally, as the bees go on to collect nectar from the following plants to bloom, the flavors of the honey will combine.

Although the color of the honey has no bearing on its quality, it does impact how much consumers think it is worth. Lighter-colored honey generally sells for more money because consumers prefer it.

The distinct floral notes of honey, often made from only one plant variety, are valued by honey connoisseurs. Gourmet honey variations with a higher value can be produced by collecting during the nectar flows of various local plants.

While you should make the most of each unique floral honey, you should also avoid harvesting too soon. Early honey harvests contain too much water and run the danger of rotting.

When the bees cap the honey cells, it is thought that the honey is both ready for harvest and at its optimum quality. One of the key elements influencing honey quality is its water content. Honey is a substance referred to as hygroscopic, which implies that it will absorb water from the atmosphere.

Keep the water content low to preserve the honey's great quality. To harvest at this period, when the weather and environment are as dry as possible. Reduce the amount of time that passes between harvest, processing, and packing.

Another way to keep the water content low and preserve the quality of the finished product is to extract, treat, and package the honey as soon as it is harvested, whether in airtight bulk containers or the final sealed consumer packaging.

Honey is typically heated to facilitate extraction and filtering and to lengthen its shelf life. The majority of commercial honey vendors favor heat-treated honey. Natural yeasts found in raw honey have the potential to become active and cause fermentation if the water content becomes too high.

The other major contributor to honey quality loss is overheating. When honey is overheated, its value is diminished because the sugars start to dissolve, and the honey darkens. Heat also eliminates the more ephemeral aromas that give honey its distinct floral characteristics, lowering its value once further.

The key to getting the most out of honey is balancing maintaining its optimal condition, preserving its unique floral features, and keeping its qualities.

Remember that you must preserve the quality of your honey if you want to sell it for the highest price. Customers will be delighted to pay a premium price for premium honey if you harvest it at the proper time, maintain the water level of your honey low, and don't overheat it.

Chapter 14: Setting Up A Beekeeping Business

There are some crucial aspects to consider and finally put into practice if someone is interested in learning how to launch a successful beekeeping business. To help you create and run such a profitable venture, I'll review some important ones.

A beekeeping business can be a wonderful way to generate a nice income and help the environment simultaneously. By their very nature, bees are natural pollinators, so they have an enormous impact on the ability of plants to regenerate and grow.

To start a successful beekeeping business, you must have a proper location, which is one of the most successful aspects of starting this business. You can't just decide to open a beekeeping business and then place bee containers in your backyard.

It's important to check with the locality where you plan to operate the business to ensure it can handle bees and bee containers. Depending on your local area and requirements, you may need permits or licenses.

Your choice of a site is very important since it greatly impacts honey's production volume and the bees' overall health. The site should be fairly flat, with some available shady areas, and it should also be located close to shrubs, flowers, or fields. The bees need sufficient areas to pollinate the plants.

Some national organizations can help you to find good locations for your beekeeping business. You can find them by searching on the Internet, and they may also be able to assist you in negotiating the purchase or lease of a site. There are also lots of beekeeping forums on the Internet.

You then need to purchase your bees. Find a qualified bee supplier early in the New Year. They are normally sold in containers that weigh between 2 and 10 pounds, and each 2-pound container holds about 8000 bees. The bees are usually shipped out during early spring (around March or April), and the carrier normally sends them out.

The bee container is the place where your bee colony resides. Plan out the number of colonies you want to maintain. In general, one beekeeper can manage about 500 to 1,000 colonies. Therefore if you want more colonies, you'll need more trained beekeepers.

The right supplies are essential to this business. You will need a queen excluder where the Queen is kept, and you'll also have several frames where the hive is formed. Protective Clothing and headgear are vital to allow you to handle the hives efficiently and safely.

Some of the most commonly used supplies for a beekeeping business include wooden frames for the hives, fume boards, feeders, queen excluders, screens, transport vehicles, brushes, a smoker (to calm the bees for transport and handling), and your protective Clothing.

The primary product which comes from a beekeeping business is honey. You will, therefore, also need equipment to assist you in extracting and processing the honey. This normally includes an extractor, sieves, and a settling tank.

Honey is a product that has strong demand. It stays fresh for a long time and is an easy product to package and sell. You can be very successful in this type of business.

Understand as much as possible about raising and handling bees and extracting and processing honey. And get the right equipment to support this. If you do so, you should be able to operate and manage a successful beekeeping business.

Chapter 15: Beeswax Projects

Beeswax is one of the most versatile substances in the world. Whether you need a colorless wax for lip balms to produce a smooth glaze on furniture, beeswax has many uses. Finding the right supplier is crucial to getting the desired quality and avoiding disappointment.

Beeswax is a natural product made by honey bees. It is used in many different industries and has many uses. The most common use of beeswax today is in candles and cosmetics, but it has also been used historically for waterproofing items such as leather goods. Beeswax is also often included in cosmetics because it contains antibacterial properties that help to prevent acne breakouts on the skin.

Salve Beeswax

Salve beeswax is a type of wax that can be used in various ways. It has a creamy, soothing texture and can be used on your skin as a moisturizer or all-natural salve. It's often mixed with other ingredients like oils and herbs to make homemade salves, lip balms, or hand creams.

Scrub Beeswax

Beeswax is a great natural abrasive, and it can be used for many projects around the house. Use it to polish wood, metal, and plastic; clean glass, windows, mirrors, and chrome; remove rust from tools; or even clean the inside of your oven!

To use beeswax as an abrasive:* Melt the wax in a double boiler on low heat.* Fill an old sock with the melted wax (this will be your applicator).* Dip your applicator into hot water; the wax should stick to it when you pull it out.* Rub over whatever surface you want to polish or clean.* Let cool before storing away again.

Body Butter Beeswax

This is a great beeswax project for the beginner. It's simple to make and requires only four ingredients: beeswax, shea butter, coconut oil, and almond oil. The beeswax can be in bars or chunks; you'll need about 8 ounces of either type for each batch of body butter. For extra soothing properties, add a few drops of vitamin E oil to your mixture before melting it over low heat on the stovetop (or double-boiler method) until it becomes clear again—this will take about 10 minutes if all of your ingredients are at room temperature. Then whip everything together until it becomes creamy smooth!

To finish off your body, butter with essential oils that complement the base notes in the oils you've chosen: orange blossom for sweet scents like vanilla bean or lavender; rosemary or sandalwood when creating fresh fragrances such as basil leaf/thyme; patchouli when making floral blends such as jasmine/ylang ylang; minty aromas like peppermint stick blend well with grapefruit/lemon grass blends etcetera...

Furniture Polish Beeswax

Beeswax furniture polish is easy to make and works on wood, metal, and plastic.

It can be used on all furniture types and is a natural alternative to commercial polishes.

Pet Care Beeswax

Beeswax and Propolis are also great for pet care. Beeswax can be used to treat:

- Skin (beeswax for skin)
- Fur (beeswax for fur)
- Ears (beeswax for ears)
- Paws (beeswax for paws)
- Teeth (beeswax for teeth).

Beeswax Lip Balm (Lip Balm, Lip Gloss)

Beeswax lip balm and beeswax lip gloss protect your lips from cold weather. The recipe is simple, requiring only two ingredients: beeswax and honey.

To begin making your own beeswax lip balm or beeswax lip gloss, you'll need the following:

- 1/4 cup of melted beeswax (you can melt at 175 degrees F)
- 1 tablespoon of honey (use raw honey)

Lotion bars beeswax

Lotion bars are the perfect way to make a moisturizing body lotion. They're easy to make, very effective, and easy to carry with you. Lotion bars are great for sensitive skin because they don't leak or spill in your purse as traditional liquid lotions do.

Fix'n Wax Beeswax

Beeswax is a wax produced by honey bees in the hive of their hives to build combs and store honey, pollen, and Propolis. The wax is secreted from glands on the abdomen of worker bees and molded into shapes by special bees called "pressers". The wax per bee varies from 1/12 oz (1g) for workers to 4/12 oz (4g) for drones.

Cleaning: To clean your candle holders, dip them into warm water with a little dish soap added. Then rinse them off with warm water again. Dry thoroughly before putting away or reusing. If you're using glass jars for storage or gift-giving (instead of votive holders), ensure all odors are out before storing anything inside them!

Making Candle Holders: Start with an old cup or bowl with better days...or buy one at a thrift shop! Paint it with acrylic craft paint if you want a fun new look, then glue pieces of fabric around the rim; cut holes in each piece so they'll sit flat atop your cup or bowl when stretched across its surface area – then secure all edges tightly with a hot glue gun - set aside until completely dry before filling up the jar with whatever goodies you'd like!

A great substitute for beeswax candles is beeswax candles. Beeswax candles can be made with beeswax, wax, and oil. Beeswax candles have a lovely scent and are used for décor.

Beeswax Food Wraps

Beeswax food wraps are made from beeswax and are used to wrap food. Beeswax food wraps keep your food fresh, preserve its flavor, and help prevent spoilage. Making your beeswax food wrap using supplies you probably already have in your kitchen is simple.

Most health food stores carry beeswax, a natural substance, and you may also buy it online. Any waxed paper can be used; cut it into squares before folding it into pockets. Use the folding method of a honeycomb here: When wrapping an object with handles, such as a bowl or container with handles for carrying outside where there are flying insects that might contaminate our foods if left uncovered during transport time, it is important to fold one corner diagonally across the square towards where you started at a 90-degree angle. Then, fold both corners on each side, meeting those diagonal edges together. Finally, fold up one end to meet the center crease.

Preserve Decorative Leaves

❖ Use a glass container with a secure lid. It also works with a tiny jar with a metal cover.

❖ Cover the jar's opening with a piece of fabric or tissue paper to keep your leaves dry. Your preserved leaf should have water added to fill the container halfway up the side. Close the container snugly, then store it in a cool location for about two weeks before removing it from the wet environment (this makes them last longer).

Conclusion

Now that you've realized you have the urge, the desire, and the chance, you can take advantage of them. How, then, would one begin keeping bees? Seasoned beekeepers compiled these eight tips and represented everything they wished they had known before purchasing their first hive.

> Use modern equipment. Naturally, used equipment is cheap or free, but it may also have problems that a novice may not notice or be able to solve. Save yourself an unnecessary migraine by starting fresh with new hives and casings.

Note: If reusing old equipment is important to you, have a seasoned beekeeper inspect it for you, and make sure you have a beekeeping teacher who can assist you should you run across problems.

> Obtain a head start on the season. To determine the best settlement time, consult beekeeping resources in your surrounding area. You don't want to start too early because the honey bees need time to find food and find a warm place to live, but you also don't want to start too late because then they won't have time to generate enough nectar for the colder months or they won't catch the first big rush of nectar.

> Simple is best. Concentrate on learning the fundamentals of honey beekeeping when you first start. Keep experimenting at the very beginning. Until you have set up sound hives, become proficient with the tried and true first.

> Run Two Checks on Your "Should Have" List. Before making any requests, be sure you know the equipment's purpose and the justification for your requirement. Everyone keeps bees uniquely, and each person has their quirks and preferences. Hold your foundational purchases to the items you must start with hives, honey bees, a smoker, and defensive gear.

› Start with Italian bees. In the US, Italian honey bees are standard, and they are typically easy to reach, are well-known by seasoned beekeepers you can ask for assistance from, and are not challenging to work with.

› With a Nucleus Colony, begin. Starting with a core province, or "bundle," of honey bees will enable you to establish a state, which will teach you a ton about raising honey bees in your specific circumstances. Although gathering wild hives is well-known at many beekeeping conferences, it is not the best option for beginning beekeepers.

› Imagine establishing two colonies. Given that two provinces require more effort than one, this may not make sense. A province to work if you lose one, typical for beginning beekeepers, and the chance to examine settlements, which will help you discover problems earlier since you'll notice a difference, are two things that two states will give you both.

› Recognize That Your First Year May Be Without Honey. It is not unusual for the newly established honey bees to fail to produce enough excess nectar in their first year, depending on your region and the weather that year. Take up beekeeping with a long-term perspective. Numerous perspectives contend that beekeeping lacks a closure game.

www.ingramcontent.com/pod-product-compliance
Lightning Source LLC
LaVergne TN
LVHW022316200926
848481LV00004B/1360